Tayna da Silva Vieira

Financial mathematics and the learning process

Tayna da Silva Vieira

Financial mathematics and the learning process

In elementary school

ScienciaScripts

Imprint
Any brand names and product names mentioned in this book are subject to trademark, brand or patent protection and are trademarks or registered trademarks of their respective holders. The use of brand names, product names, common names, trade names, product descriptions etc. even without a particular marking in this work is in no way to be construed to mean that such names may be regarded as unrestricted in respect of trademark and brand protection legislation and could thus be used by anyone.

Cover image: www.ingimage.com

This book is a translation from the original published under ISBN 978-613-9-61429-5.

Publisher:
Sciencia Scripts
is a trademark of
Dodo Books Indian Ocean Ltd. and OmniScriptum S.R.L publishing group

120 High Road, East Finchley, London, N2 9ED, United Kingdom
Str. Armeneasca 28/1, office 1, Chisinau MD-2012, Republic of Moldova, Europe
Printed at: see last page
ISBN: 978-620-7-62866-7

Table of contents:

Foreword

This book is aimed at anyone with an interest in learning Financial Mathematics, especially teachers and students. But for readers who want to plan their investments well and develop objectives that involve Financial Mathematics, reading this book will be very beneficial.

The author

THANKS

Thank you for the excellent support received from the Novas Edipóes Académicas editorial team. I would also like to thank all those who have been and are close to me in some way, offering total support in my professional life.

Chapter 1

1. INTRODUCTION

The transformations in the Brazilian economic scenario over the last decade have seen a major decline, aggravated by the current crisis. It is true that the economic downturn did not happen suddenly. Faced with such a scenario, it is necessary to explore the study of Financial Mathematics so that the population is prepared and knows how to deal with issues related to financing, indebtedness, the value of money over time, consumption, pensions, etc.

However, despite the crisis, we live in a capitalist system where the main objective is profit. As such, companies invest heavily so that, regardless of the crisis, they can sell. As a result, people are bombarded with marketing strategies to turn them into consumers, especially those consumers who act on impulse when buying, without knowing how to properly analyze whether the product or service is necessary or superfluous, and without coherently analyzing the payment methods offered by the market. Not to mention the influence of the media and the ease of obtaining credit from banks. According to Skovsmose (2005, p. 104),

> "Nowadays, we are witnessing the insertion of

technological products in different areas of society, whether at home, at work, during leisure or consumption. Therefore, reflecting on the presence of mathematical concepts in these products also represents a reflection on how mathematics is present in routine activities. Consequently, in order to assess the influence of certain mathematical concepts in our lives, we need to develop the ability to recognize them in our daily lives."

However, educating society financially goes beyond teaching financial mathematics, despite its fundamental importance. The aim is broader: to make people act with focus, to know how to make decisions and diagnose the best purchasing option. And it is with this aim in mind, of better teaching Financial Mathematics, that an analysis will be made of the teaching material offered in Primary School.

In this sense, it is important to emphasize the importance of the learning environment for teaching financial mathematics, since today's students will be the future citizens who will have to deal with financial issues in their adult lives. It is therefore constructive to bring exercises with real information into the classroom, inviting the student to discover the best financial way to act in the future. It is also important to take into account the students' prior knowledge, which is usually overlooked, helping to make lessons less mechanical and without schematic calculations.

In a multidisciplinary context, connections can be made beyond the economic, such as working on students' social issues (labor relations, economic diversity, citizenship, etc.), environmental issues, making students reflect on the environmental impacts of unbridled consumption and behavioral issues (desire versus need).

If we understand that the teaching of mathematics contributes to the empowerment of the individual in society, we will come to the conclusion that approaching financial mathematics in elementary school is an extraordinary opportunity for this. For this reason, this work seeks to analyze the teaching material used in elementary school for the final grades, in order to see if it is suitable for such teaching. If it isn't, find out why and try to find tools to improve it.

Chapter 2

2. OBJECTIVES

Some interesting questions guided the objectives of this work.

Is Brazilian society adequately educated on the subject of financial education?

What materials are used to teach financial mathematics in elementary school for the final grades?

How is this knowledge applied in schools?

How are teachers prepared to teach a class on this subject? And does the school environment influence the learning of Financial Mathematics? What impact does this teaching have on the economic scenario?

Special attention should be paid to Financial Education. To do this, we need to make sure that the school develops this teaching in the most appropriate way, using all the means to teach Financial Mathematics. It is already clear that the government, society and those directly involved in the economic world, such as banks, companies and bodies like the Central Bank, are aware of the importance of this subject. And the school environment is a medium that reaches a significant portion of the population. Now, if schools manage to develop critical, quality math education, Brazil will have future citizens with a greater capacity for financial analysis, and this is fundamental for the country's economic development.

According to Fonseca (2005, p. 83),

> "Critical Mathematics Education provides social empowerment by developing mathematical literacy or social numeracy, allowing the subjects involved to exercise critical judgment in political and social decisions."

2.1 General Objective

In order to verify and help the country's economic development and believing that having a more critical population with a better theoretical foundation in Financial Education will help in this development, this work will include an analysis of textbooks used in Primary School for the final grades and field research in a school in the state of Rio de Janeiro, to check the methodology applied in Primary School with regard to Financial Education.

It is necessary to open up the range of students' learning, providing a greater perception of the problem, making students take a critical look at the subject, not allowing knowledge to be pruned for them. According to Kistemann

(2012, P.7)

> "With an understanding of the mathematical concepts that are present in every situation in this society, each individual will be able, through their mathematical knowledge and by making critical readings of situations using this knowledge (Mathematics), to make their decisions based no longer on the hegemonic directives of a state nobility, but on their critical mathematical knowledge. ".

2.2 Specific Objective

To analyze the teaching materials used in elementary school, taking as a source of analysis a state school in Rio de Janeiro, and to verify their effectiveness in student learning. In addition, to discuss the impact of the teaching of Financial Education on Brazilian society. Field research will be used as a critical parameter. The main issue here is to find out how this subject is taught in schools, as well as proposing alternatives to teaching the content in the final years of elementary school.

Chapter 3

3. METHODOLOGY

3.1 Historical Perspectives

To begin the development of the work, we will take a trip back in time to see part of the origin and evolution of money and thus be able to make a better argument about current financial education in Brazil.

When man stopped being nomadic and started living in a fixed place, the first idea of trade began to emerge. The economy became based on the exchange of goods, an activity known as barter. This was the form of trade that dominated at the beginning of civilization.

However, over time there were problems, as the exchange of some goods for others was not always advantageous. Salt and cattle, for example, gained greater prestige and became commodity currency. And this is where the words that we still use in our vocabulary today came from: pecunia (money) and peculio (accumulated money), derived from the Latin word pecus (cattle). Similarly, the word salary comes from the use of salt.

With the discovery of metal, it became the main standard of value due to its ease of transportation, hoarding, beauty, etc. Later, it took on forms such as knives and/or keys. In the 7th century BC, the first coins were minted with characteristics similar to those we have in circulation today. The Middle Ages saw the emergence of paper money, which at first was a receipt for valuables kept with the goldsmiths.[1]

This makes us reflect on how man needs to adapt in order to develop as a society.

It's clear to see that major changes have taken place over time. The Modern Age is the time when the capitalist system began to be introduced and it was from there that the idea of profit, purchasing power, interest, etc. began to emerge.

3.2 Specific analysis: Textbook Guide 2014, Elementary School - final years

Much of what is taught in schools today, in relation to Financial Mathematics, is guided by textbooks that are often used without questioning their applicability and suitability to the different realities of the students. How do the textbooks reach the students? What analysis is made of their choice? Who does this analysis?

---

[1] Information taken from the website of the Central Bank of Brazil at http://www.bcb.gov.br/htms/origevol.asp?idpai=HISTDIN, in March 2016.

The purpose of the National Textbook Program (PNLD) is to distribute textbooks to students, thus helping the work of teachers. The books currently used in schools for the final years of elementary school were chosen in 2014 with the aim of using them for three years. Before they are distributed to the students, the Ministry of Education (MEC) examines them and publishes the Textbook Guide with a summary of the collections chosen, after which the schools choose the textbook they prefer and which meets their pedagogical policy. The process of choosing textbooks is the responsibility of the teachers and school leaders. For each curricular component, two choices of books from different publishers can be made, because if there is a shortage of the first choice, the second choice is sent instead.

All the information about this choice is available on the portal www.fnde.gov.br for the guidance of schools and teachers. Just go to "programs" and PNLD.

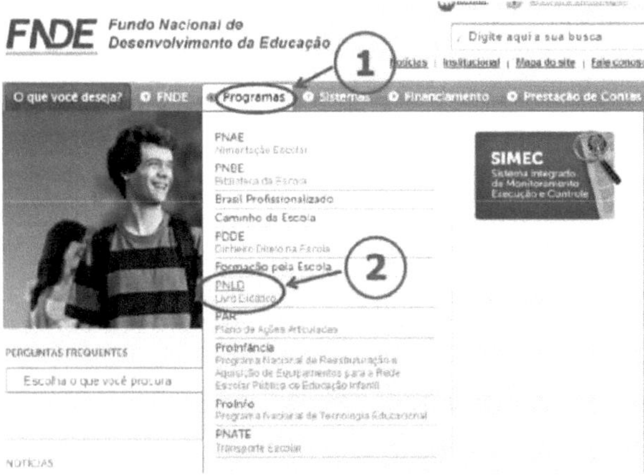

Figure 1 Guided access to the portal for choosing textbooks Source:

http://www.fnde.gov.br/programas/livro-didatico/guias-do-

pnld/item/4661-guia-pnld-

2014

Figure 2 Guided access to the portal for choosing textbooks Source:

http://www.fnde.gov.br/programas/livro-didatico/guias-do-pnld/item/4661-guia-pnld-

2014

To actually choose the collection you want, just go to the system and click

on SIMAD, which is the Didactic Material System, and proceed with access

using a username and password.

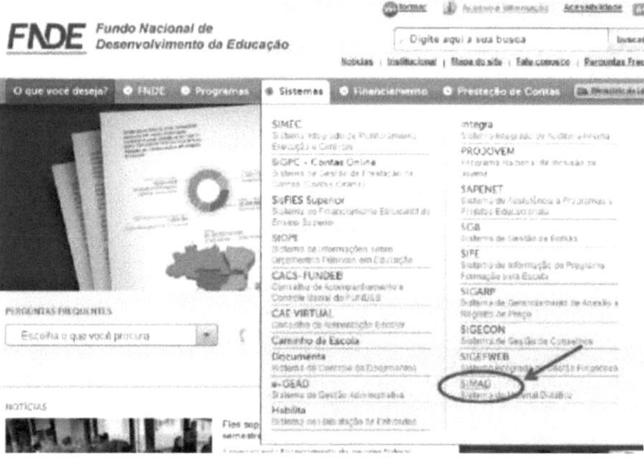

Figure 3 Guidance for accessing the portal for choosing textbooks Source:

http://www.fnde.gov.br/programas/livro-didatico/guias-do-

pnld/item/4661-guia-pnld-2014

Ascertaining that Ciep 355 Roquete Pinto uses the Bianchini Mathematics

collection from the 6th to the 9th year of elementary school, and that the

books for the 7th and 9th year of elementary school are the books that are

closest to the content of Financial Mathematics, the two books were analyzed and it was checked whether they are in fact in accordance with the PNLD 2014 Textbook Guide and, most importantly, whether they are developing the student's critical sense so that they know how to deal with the issues of their daily lives.

The textbook is a material whose purpose is to help the student learn, being a mediator between teacher and student. And the National Textbook Plan aims to provide detailed information so that a good choice of textbooks can be made, in accordance with each school's political and pedagogical project.

3.3 Textbooks: conceptualization and approach

The MATHEMATICS BIANCHINI Collection by Edwaldo Roque Bianchini for the final years of elementary school has the following overview, according to the Book Guide Didáticos 2014, p.17, that:

> "The content is covered by explaining the theory, accompanied by examples and the Proposed Exercises section, which presents problems for applying what has been taught. In general, this methodology doesn't give the student much opportunity to develop the knowledge to be

acquired in a more autonomous way. Despite this, situations are proposed in which the student's ability to argue is mobilized to justify their solution strategies and their answers. Some more thought-provoking problems are further opportunities for students to exercise their creativity. "

However, the methodology needs to give students the opportunity to exercise their authenticity, giving them the chance to share their previous knowledge, their experience in the classroom, as well as being motivated to build other knowledge that they haven't achieved so far. In this way, the mutual sharing of learning between teacher and student takes place. After all, teacher/student interaction is a condition of the learning process. "*Those who teach learn by teaching and those who learn teach by learning*" (Freire, 1996, p.38).

The topic Teaching and Learning Methodology reinforces this idea, and also emphasizes the use of calculators and the limited use of games and technological resources. One fact that should be highlighted is that the use of electronic games, simulators, videos or infographics (called Digital Educational Objects) is a new feature in the 2014 PNLD to innovate and improve classroom work. In addition, he claims that it seeks to systematize the content and thus depreciates the knowledge and experience that each student has outside the classroom. The 2014 Textbook Guide, p.19 states that:

> "The methodology adopted begins with explanations and examples which seek to systematize the content to be studied, followed by the Proposed Exercises section. Generally, there are few opportunities for students to develop concepts and procedures more autonomously and to link them to others they have already learned. In addition, students' extra-curricular knowledge is not highly valued. (...) Concrete materials are used adequately as teaching aids. The use of calculators is frequent, unlike games and technological resources, which are not very common."

Figure 4 Cole^áo Matemática Bianchini by Edwaldo Roque Bianchini

Source: PNLD 2014 Textbook Guide

The theory of Ausubel et al. (1981) deals specifically with the processes of teaching and learning scientific concepts based on concepts previously formed by students in their daily lives (Pozo, 1989). Ausubel (apud Novak, 1981, p.9) states that: *"the most important single factor influencing learning is what the learner already knows. Determine this and teach accordingly"*. Based on this line of thinking, the school must take into account the student's knowledge and, more than that, the teacher needs to know how to exploit this knowledge so that they can teach according to what the student already knows. Not necessarily being "tied down" to the ideas set out in the textbook. After all, the textbook is just an aid to the lesson. The contents of the BIANCHINI MATHEMATICS collection are classified as follows.

See figure below.

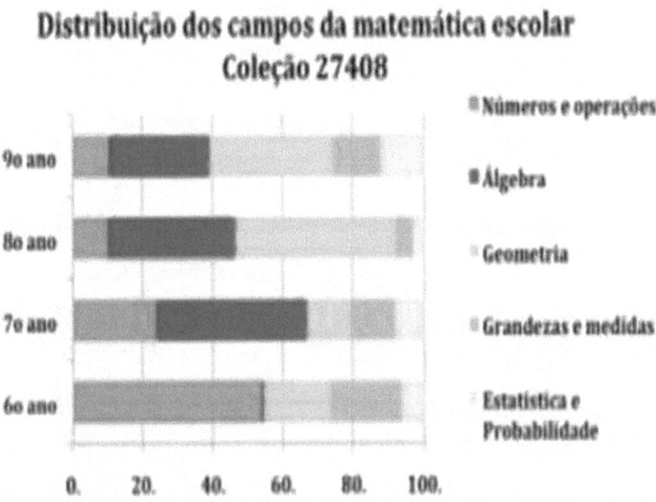

Figure 5 Distribution of content in the Bianchini Mathematics collection

Source: PNLD 2014 Textbook Guide

In none of the topics: Numbers and Operations, Algebra, Geometry, Quantities and Measures, Statistics and Probability, is Financial Mathematics actually covered. Monetary value is a quantity mentioned in the topic of Quantities and Measures, but the focus is on geometric quantities.

The contents of Statistics and Probability are developed a little more in the ninth year of elementary school, being developed in a single chapter. In the Guide to Textbooks 2014, p.19, with regard to the selection and distribution of content, it says in Statistics and Probability that *"A caveat should be made in the study of measures of central tendency, which are approached quickly, without much discussion between the calculations and their interpretation in the context in which they are inserted. "*

As a result, we realize that Financial Mathematics has not been approached in an effective way in Primary School, just by looking at how superficially this subject is treated in the textbooks used.

In addition, we need to understand that we shouldn't just teach concepts, but that we should build knowledge. D'Ambrósio (1996, p.80) states that: "*The great challenge for education is to put into practice today what will serve tomorrow*".

3.4 **Field research**

A survey was carried out with the aim of investigating the extent to which Financial Mathematics has been addressed in students' lives, both at school and outside it, and how significant it has become in students' daily lives.

Fifty-seven students aged between 14 and 17 from the eighth and ninth grades at CIEP 355 Roquete Pinto, located at Estrada Rio Douro, E/F 2000, s/n - Paraíso, Queimados - RJ, 26382-210. Research authorized by the Director of CIEP.

The survey consisted of eight questions[2] . These are as follows:

Question 1 - Have your parents and/or guardians ever talked to you about buying, saving, economizing, researching prices or anything like that?

( ) Yes

( ) No

Question 2 - Have your teachers ever talked to you about buying, saving, researching prices or anything like that? (Apart from percentage accounts).

( ) Yes

( ) No

Question 3 - Mark the contents that you remember having studied in previous years.

( ) Percentage

( ) Ratio and Proportions

( ) Proportional quantities

---

[2] Five questions taken and one adapted from the poster: Proportionality in consumer decision-making. Presented at the XII National Meeting of Mathematical Education by the Geedufin group.

( ) Statistics and Probability

Question 4 - Do you use the math you've learned to decide "what" and "how" to buy things?

( ) Yes, often.

( ) Yes, rarely.

( ) No

Question 5 - An ice cream store offers two types of jars:

I - 200 ml for R$ 5,00

II - 350 ml for R$10,00

Which of these options is the most interesting in financial terms?

Justify!

Question 6 - Look at the packaging of a famous cookie:

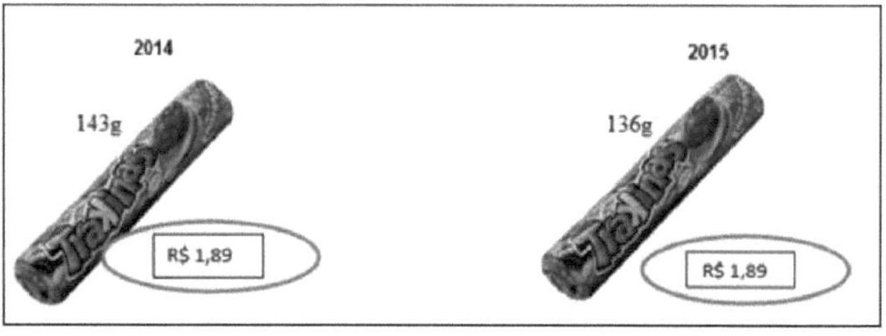

Figure 6 Relating to question 6 of the field survey

From 2014 to 2015, according to the information provided, the relative price of the cookie. Justify:

(   ) increased.

(   ) decreased.

(   ) has been maintained.

Question 7 - Have the math textbooks you've used helped you learn the concepts of Financial Mathematics?

( ) Yes

( ) No

Question 8 - How do you rate the quality of the textbook when it comes to Financial Mathematics?

( ) Excellent

( ) Good

( ) Fair

( ) Bad

( ) Bad

3.5 Analysis of results

With regard to the first question, the aim is to analyze whether students are guided not only at school, but also by their parents and/or guardians.

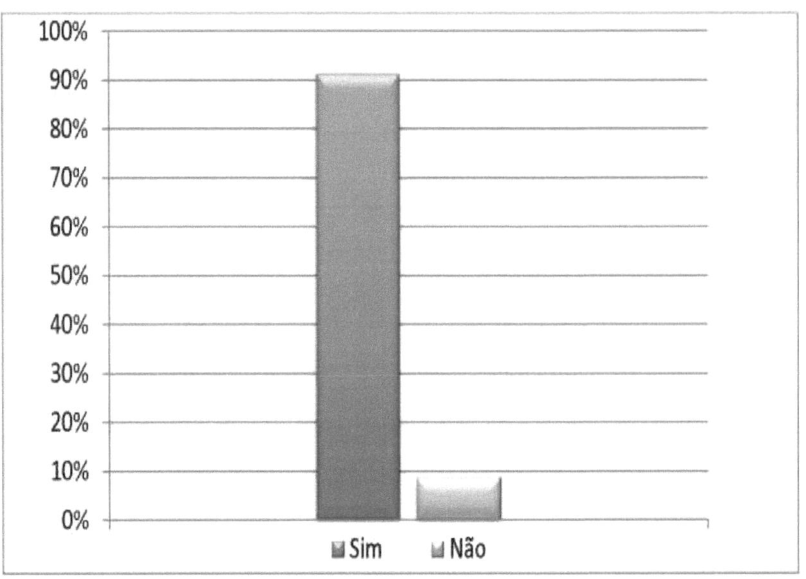

Figure 7: Graph of the results of question 1

It is worth noting in these results that almost all students (91%) are guided by their parents and/or guardians, which is a very positive result. However, there is a possibility that the student may have given an answer that is more to the liking of their parents/guardians and is not a complete representation of the truth. Even if they knew that their identity would not be revealed.

In the second question about teacher guidance, the answers were different from the first question: 37% of the students said that yes, they do receive guidance from their teachers, while 63% said that no, they are not guided in this direction by their teachers. This result raises alarm bells about the quality of math education being offered to these students.

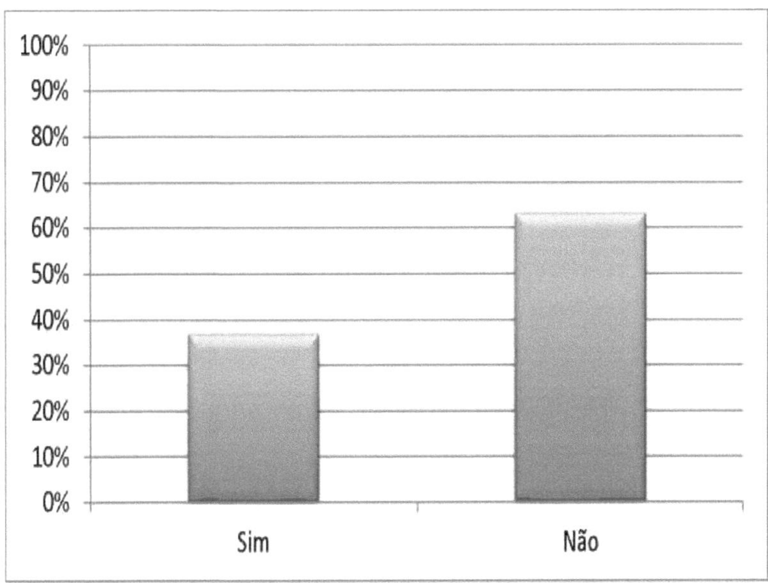

Figure 8 Graph of the results of question 2

The third question shows worrying results about the knowledge that students have or should have about mathematical subjects related to Financial Education. Students have a good knowledge of percentages, but when it comes to Ratios and Proportions, Proportional Magnitudes and Statistics and Probability, few students remember the subject. There are still 2% of students who don't remember any of the content.

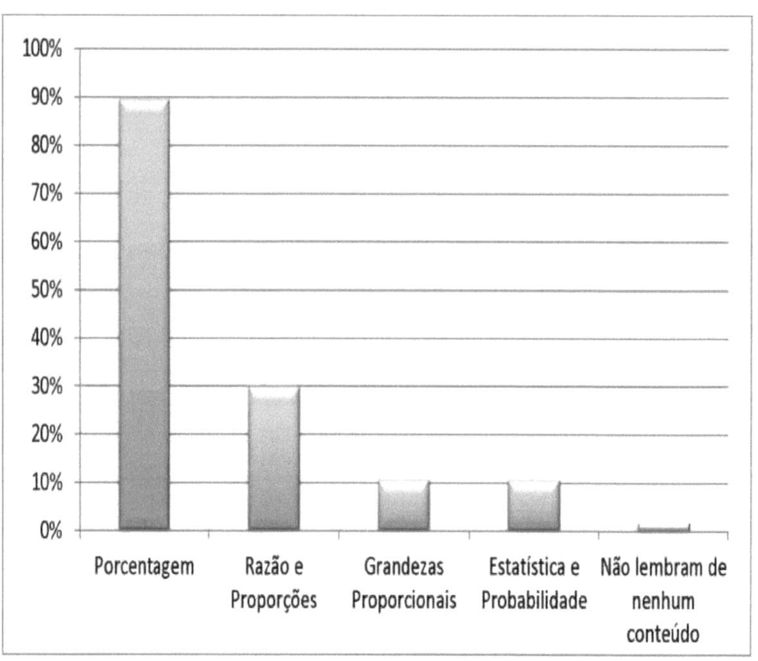

Figure 9 Graph of the results of question 3

In the fourth question, the results were that 44% of the students use math when making decisions to buy, and that 39% use math, but rarely. What is worrying about this question is that the other 18% of students say they don't use mathematics when making decisions to buy, which shows us that they could be future citizens who act impulsively when buying.

For Skovsmose (2002, p. 78),

> "It is of the utmost democratic importance, both for the individual and for society as a whole, that any citizen has at their disposal the tools to understand the role of

mathematics (in society). Anyone who doesn't have these tools becomes a "victim" of the social processes of which mathematics is a component. Thus, I believe that the aim of Mathematics Education should be to enable individuals to perceive, understand, judge, use and also apply Mathematics in society, especially in situations that are significant for their private, professional and social lives. Mathematical literacy, as a radical construct, must be rooted in a spirit of criticism and in a project of possibilities that enables people to participate in the understanding and transformation of their society. "

Following Skovsmose's line of thought, there needs to be a change in this situation, so that all students can apply mathematics in their social and professional lives.

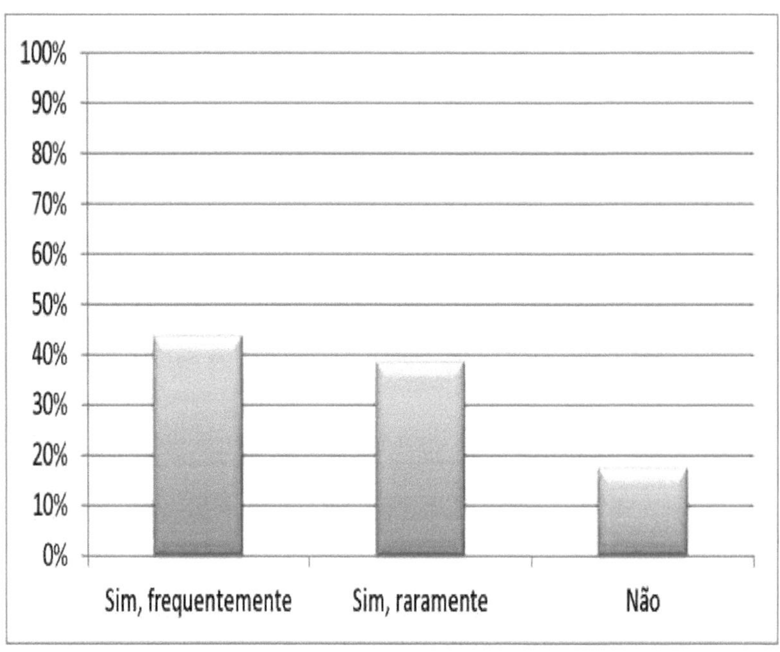

Figure 10 Graph of the results of question 4

In the fifth question involving Proportional Quantities, content that only

11% of the students said they remembered from the third question, 63% of

the students understood that the 200ml cup was proportionally more

useful. 21% said that the 350ml cup was the best choice and 16% couldn't

answer. This result reflects and consolidates the previous ones.

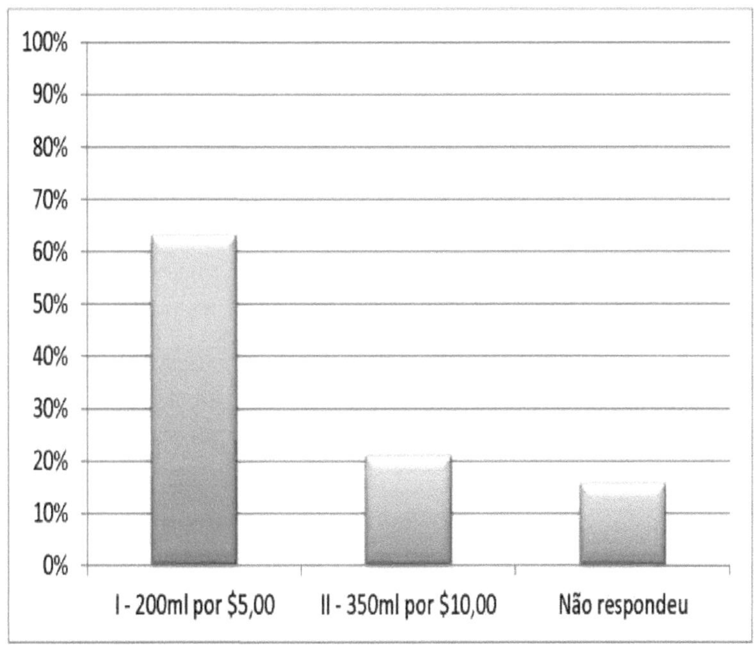

Figure 11 Graph of the results of question 5

The sixth question showed a very poor result, as only 11% of the students realized that the price had increased. A further 23% of the students said that the price had fallen, which worsens the picture. And 67% said that the price had remained the same.

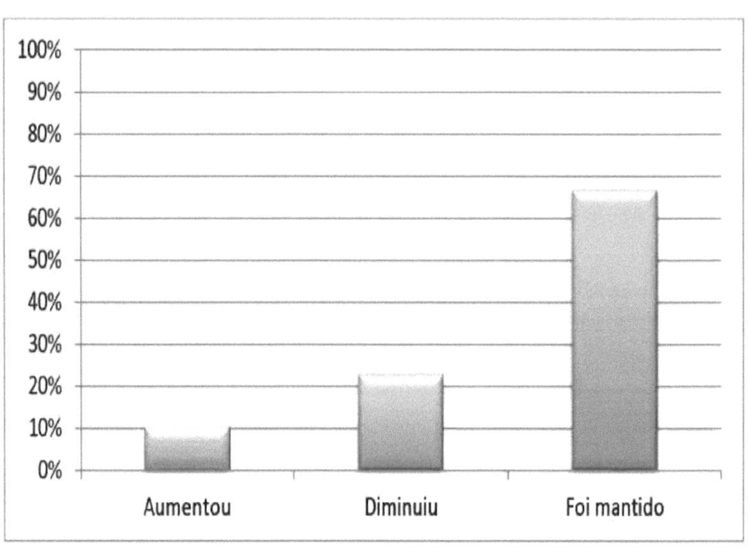

Figure 12 Graph of the results of question 6

With regard to the majority of students who said that the value was maintained, it is possible to see from their answers that they observed that the amount in grams of the packet of cookies had decreased, however, they were unable to see that decreasing the amount of the packet of cookies would mean that the value of the cookie had increased, leading to the conclusion that the concept of proportional quantities was not fully understood by these students. See the image below for one of the students' answers:

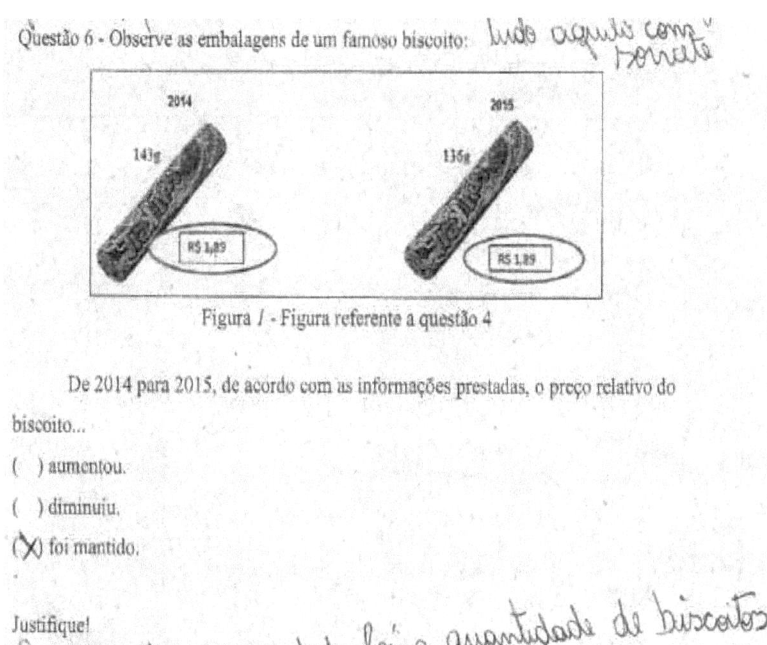

Figura 1 - Figura referente a questão 4

Figure 13 Student (a)'s answer to question 6

In the seventh question of the survey, a question about textbooks and the support they offer students on the concepts of Financial Mathematics, the following result was obtained: 61% of the students believe that yes, textbooks provide the expected support with regard to Financial Mathematics and 39% believe that no, they do not provide the ideal support. This proves to be a contradiction, given the results obtained in previous questions.

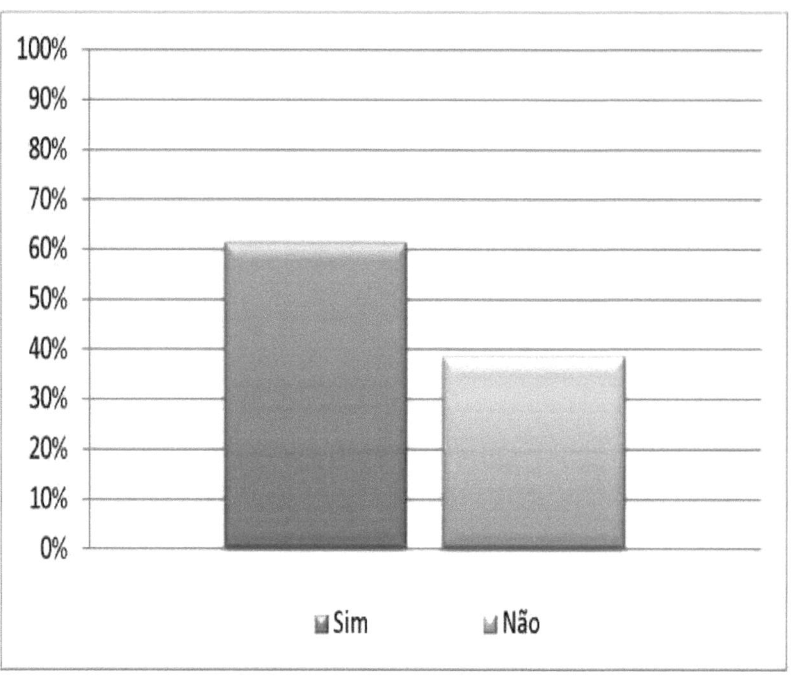

Figure 14 Graph of the results of question 7

The eighth question asked the students themselves to evaluate their textbooks. In the opinion of the majority of students (47%), the textbook when it comes to Financial Mathematics is reasonable, 11% consider it excellent, 33% good, 5% bad, 2% terrible and 2% couldn't answer.

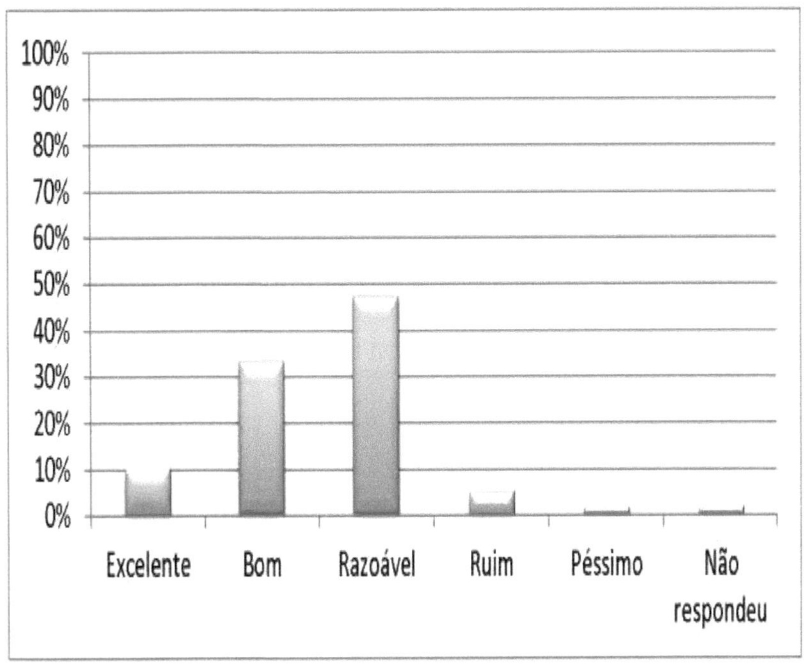

Figure 15 Graph of the results of question 8

Chapter 4

4. A PROPOSAL FOR CRITICAL FINANCIAL EDUCATION.

All Brazilian families have dreams and projects to fulfill, but society needs to mature financially and know how to make conscious use of the money and financial products on the market. On the other hand, society is currently focused on consumerism and the devaluation of consumer goods. Today's coveted electronic gadgets will be junk in the near future.

According to Bauman (1999 p.88):

> "   Our society is a consumer society .
>
> When we talk about a consumer society, we have in mind something more than the trivial observation that all members of that society consume; all human beings, or rather all living creatures, have "consumed" since time immemorial. What we have in mind is that ours is a "consumer society" in the similarly profound and fundamental sense that the society of our predecessors, modern society in its founding layers, in its industrial phase, was a "producer society. [...] The consumer in a consumer society is a markedly different creature from the

consumers of any other society up to now. "

In order to form critical citizens who are accustomed to making coherent decisions on the subject of Financial Education, it is necessary to explore open questions from an early age in schools, such as discussing the best form of payment and whether "buy two get three" promotions are really advantageous. Making students have contact with this universe that awaits them in the future, after all dealing with Financial Mathematics is exactly that: acting in the present with a view to the future.

For students, it is more desirable to learn from circumstances that involve their daily lives or cases in which they can interact. To meet this premise, teachers can favor active student participation. The idea discussed here is not intended to prescribe the methodology that the teacher will use in his or her class, but only to open up the range of possibilities in order to broaden the student's horizon. Ideally, an analysis and reflection of consumer situations should be promoted first, always remembering the student's prior knowledge so that learning becomes meaningful. Below are some examples of how some thought-provoking questions can lead to richer discussions from the point of view of critical financial education.

1) Maria wants to buy a blouse. She knows that her favorite store offers its

customers two payment methods: cash or in two equal installments. The blouse Maria wants is on sale for a total price of R$200.00 for payment in two installments, R$100.00 at the time of purchase and R$100.00 30 days later. Which of the two options is the best way to buy it?

ANSWER: The question prompts reflection, as there are different points to be analyzed. If Maria knows how to save money, she can choose to buy in two installments without any problem, or better still, Maria could invest the remaining R$100.00 to save some money for the following month. Another possible option is to buy in cash, which is suitable for impulsive consumers who don't know how to save, and Maria could also ask for a discount.

2) A plastic jar of ice cream costs R$20.00 for one liter. A certain market sells the same ice cream in a plastic container for R$40.00 for two liters. Which option is more advantageous: two one-liter jars of ice cream or one two-liter jar of ice cream?

ANSWER: In question 2, there are actually no financial savings to be made. However, you can take the environmental issue into account, making it convenient to say that buying the biggest jar will do less damage to the environment, since more plastic is used in making two one-liter jars than in one jar that holds two liters of ice cream.

3) Thiago went for a snack with his sister Ester, his favorite snack cost R$20.00 and his sister's favorite snack cost R$15.00. When he got to the snack bar, he saw that there was a "two for one" promotion where two snacks would cost R$25.00, but they weren't his favorite snacks.

a) If you were Thiago, would you buy the promotion or opt for your favorite snacks?

b) In this case, is the "two for one" promotion advantageous? Why? Explain using the calculations.

ANSWER:

a) The answer to letter a is a personal one, but it implies analyzing whether the student in question tends to make thoughtful decisions in order to analyze their financial conditions or whether they prefer to fulfill their desires and consume what they like best.

b) The "take two pay one" promotion is advantageous because Thiago would save R$10.00, money that Thiago could invest and make profitable.

c) Pedro receives R$150.00 in pocket money from his father and has to pay for all of his expenses, such as his ticket to school, lunch, outings, etc. As time goes by, Pedro notices that R$30 a month is left over from his allowance and he wants to buy a new video game that costs R$180.00.

a) How long will Pedro have to save up money to buy the game?

b) If Pedro uses his father's card and pays for his game in 6 installments of R$35.00, will it be more advantageous? What other way can he buy the game?

ANSWER:

a) Pedro will need to save his allowance for 6 months if he keeps the money in a safe and doesn't invest it.

b) 6 x R$35.00= R$210.00. Minus the original value of the video game, R$210.00 - R$180. 00=R$30.00   . Pedro will pay $30.00 in interest. Therefore, it won't be advantageous, because Peter will pay $30.00 more. A good way for Peter to buy is to save his pocket money until he has the amount needed to buy the video game, or to invest it so that it pays off and he can get the money to buy the video game even faster.

The sample questions can stimulate the production of meaning on the part of the students, making them reflect, leading them to think about the near future, where they will be more active consumers and will need to make good decisions when making purchases. Question four also conveys the idea of saving part of their income to make a purchase in the future.

FINAL CONSIDERATIONS

Educating people to consume and save in an ethical, conscious and responsible way. Avoiding the growth of default and indebtedness among Brazilian families. These are the great benefits for society once its citizens are literate in issues involving financial mathematics. By failing to address financial mathematics in the final years of elementary school, it can end up becoming difficult in secondary school and, above all, meaningless. This possible lack of interest on the part of the students is due to the late way in which Financial Mathematics is presented. We have seen that many important aspects have been left out, especially the bringing together of the student's reality with the teaching of the subject.

It is only in secondary school that the content begins to be worked on, which is contradictory to what we have seen in this work, given that students experience and are surrounded by issues involving the financial world. Therefore, if they are not developed together, the building of mathematical knowledge will not have a solid foundation.

It can also be seen that the textbooks are not fully meeting the needs of the students and, because they have a generalist character, they are far from what would be ideal with regard to Financial Mathematics.

BIBLIOGRAPHICAL REFERENCES

AUSUBEL, D., NOVAK, J. D., & HANESIAN, H. **Psicologia Educacional**. Rio de Janeiro: Editora Interamericana, 1980.

BANCO CENTRAL DO BRASIL. **Museu de Valores** do **Banco Central** Available at: <http://www.bcb.gov.br/htms/origevol.asp?idpai=HISTDIN>. Accessed on March 3, 2016.

BAUMAN, Zygmunt. **Globalization: the human consequences**. Rio de Janeiro: Editora Zahar, 1999.

D'AMBROSIO, Ubiratan. **Mathematical Education: from theory to practice**. Papirus Editora, 1996.

FONSECA, Maria da Conceiçao F. R. **Educaçao Matemática de Jovens e Adultos: especificidades, desafios e contributes**. Belo Horizonte: Editora Autêntica, 2005.

FREIRE, P. **Pedagogia da autonomia: saberes necessárias à prática educativa.** Sao Paulo: Editora Paz e Terra, 1996.

NATIONAL EDUCATION DEVELOPMENT FUND. **PNLD**

2014 textbook guide - Final Years of Primary Education.
Available at: <http://www.fnde.gov.br/programas/livro-didatico/guias-do-pnld/item/4661-guia-pnld-2014>. Accessed on February 12, 2016.

GEEDUFIN. **Proportionality in consumer decision-making**. XII ENEM. Sao Paulo, 2016. (Poster)

KISTEMANN, M. A. **On the production of meanings and decision-making by individual consumers**. V International Seminar on Research in Mathematics Education. Petrópolis, Rio de Janeiro, Brazil, 2012.

MINISTRY OF EDUCATION. **The National Textbook Program** Available at: <http://portal.mec.gov.br/pnld/apresentacao> Accessed on February 2, 2016.

NOVAK, J. D. **A theory of education**. Sao Paulo: Editora Pioneira, 1981.

POZO, J. I. **Teorías cognitivas del aprendizaje**. Madrid: Editora Morata, 1989.

SKOVSMOSE, Ole. **Foregrounds and Politics of Learning Obstacles. For the Learning of Mathematics**, 25(1), 4-10, 2005.

SKOVSMOSE, Ole. **Mathematical Education and Democracy.** *Educational* **Studies in Mathematics**, 21, 109-128, 2002.

I want morebooks!

Buy your books fast and straightforward online - at one of world's fastest growing online book stores! Environmentally sound due to Print-on-Demand technologies.

Buy your books online at
www.morebooks.shop

Kaufen Sie Ihre Bücher schnell und unkompliziert online – auf einer der am schnellsten wachsenden Buchhandelsplattformen weltweit! Dank Print-On-Demand umwelt- und ressourcenschonend produziert.

Bücher schneller online kaufen
www.morebooks.shop

Printed by Books on Demand GmbH, Norderstedt / Germany